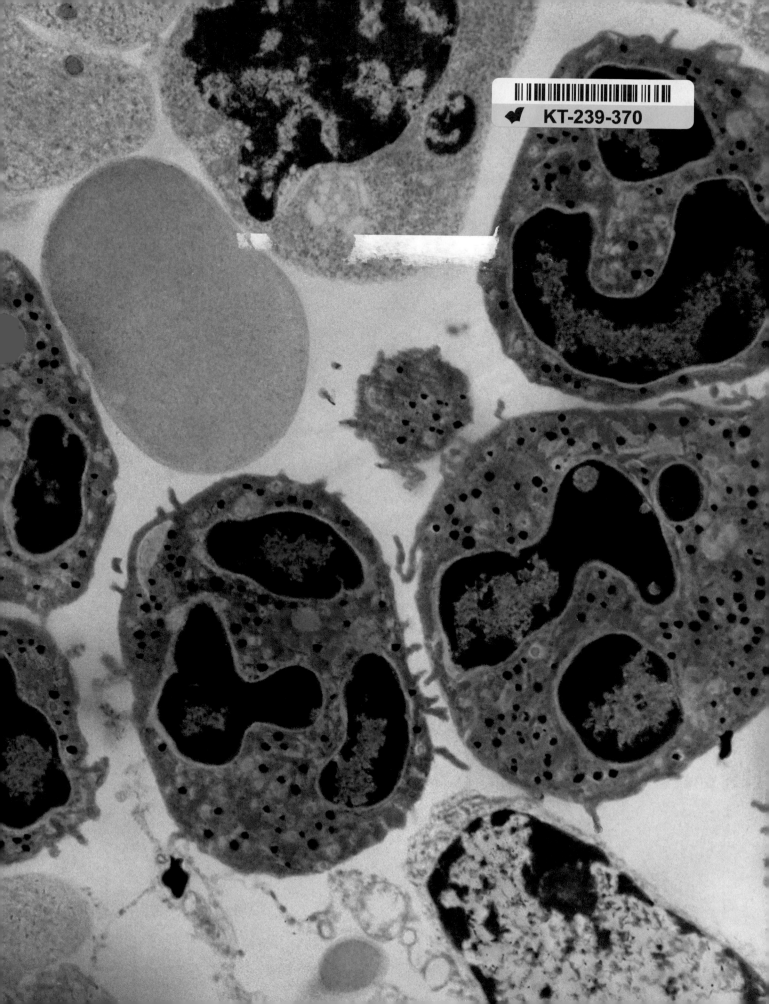

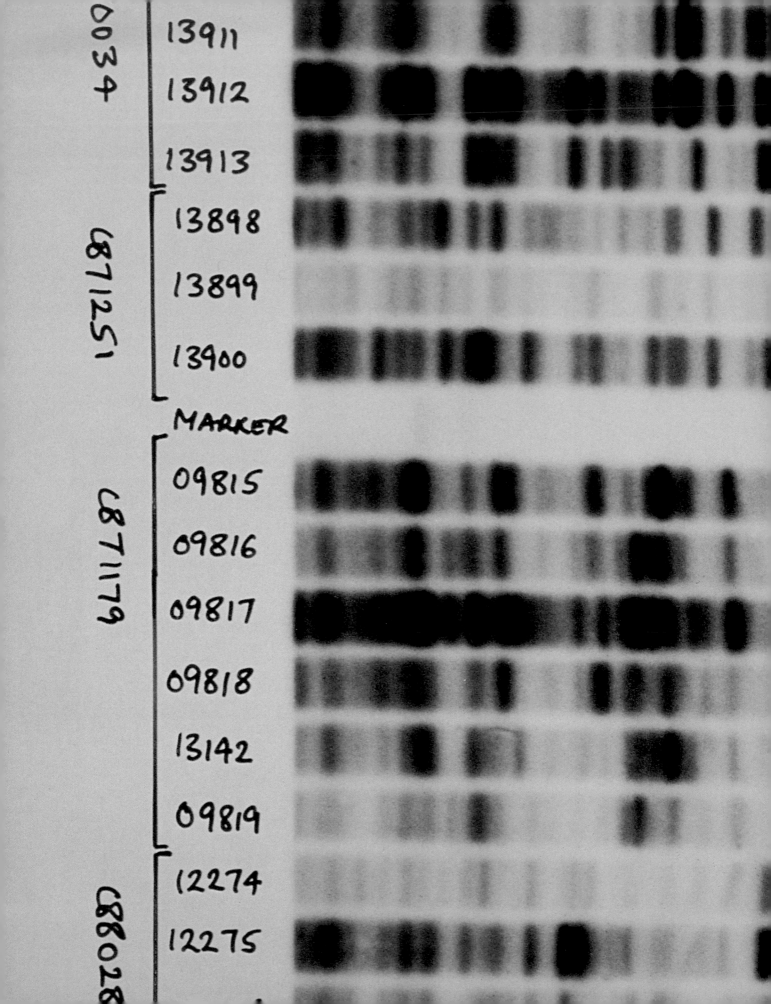

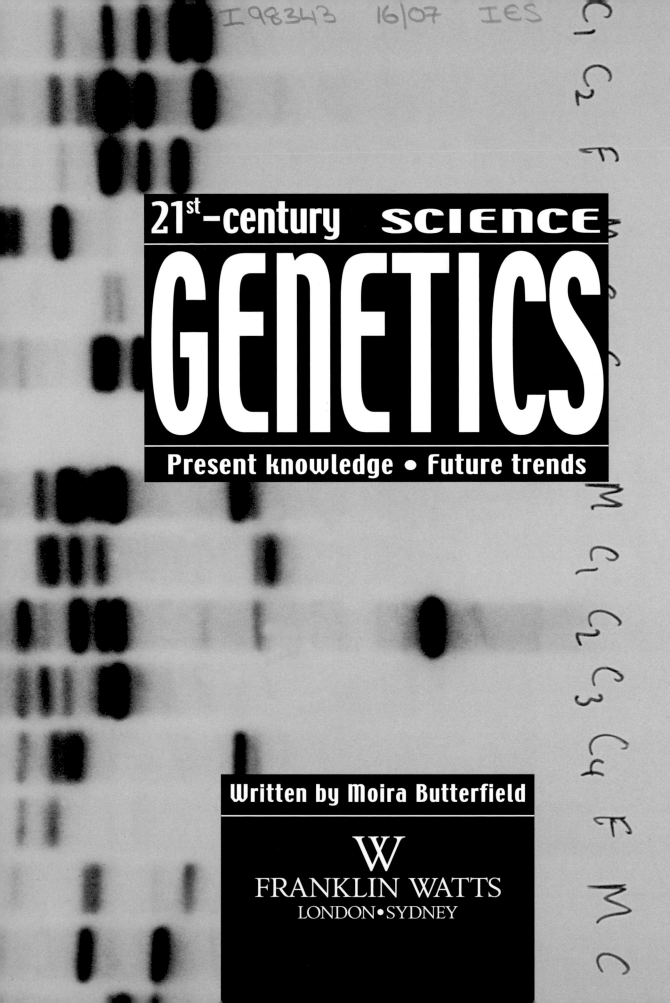

21st-century SCIENCE
GENETICS
Present knowledge • Future trends

Written by Moira Butterfield

W
FRANKLIN WATTS
LONDON•SYDNEY

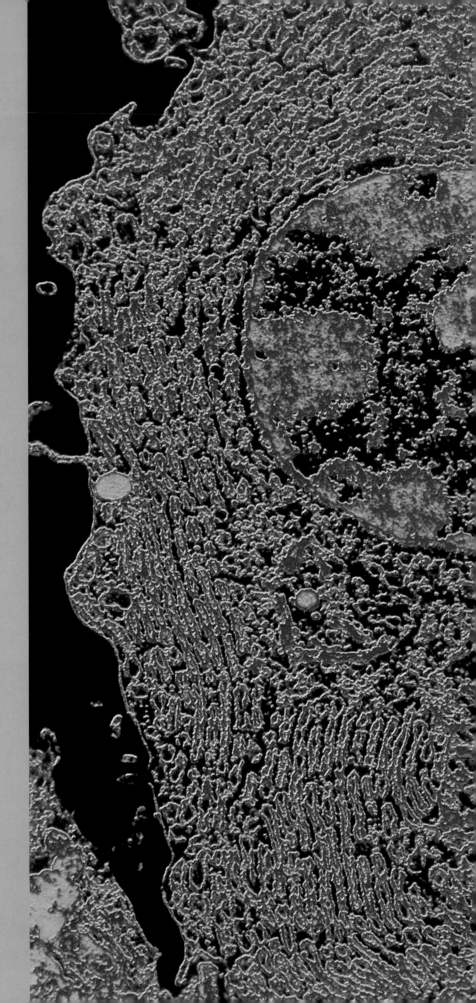

First published in 2002 by

Franklin Watts
96 Leonard Street
London EC2A 4XD

Franklin Watts Australia
45-51 Huntley Street
Alexandria
NSW 2015

© Franklin Watts 2002

Artwork Garry Walton
Design Billin Design Solutions
Editor in Chief John C. Miles
Art Director Jonathan Hair
Picture research Diana Morris

Consultant Dr Ruth Newbury-Ecob
Consultant Clinical Geneticist,
St Michael's Hospital, Bristol

A CIP catalogue record for this book
is available from the British Library

ISBN 0 7496 4587 3

Printed in Hong Kong, China

Picture credits
Alex Bartel/SPL: 20.
Biofoto Associates/SPL: 10, 11.
Department of Clinical Cytogenetics,
Addenbrookes Hospital/SPL: 18.
Eye of Science/SPL: 31.
Mauro Fermariello/SPL: 29.
Michael Gilbert/SPL: 35.
James King-Holmes/SPL: 37.
Chris Knapton/SPL: 39.
Kwangshin Kim/SPL: 6.
Francis Leroy/Biocosmos/SPL: 13
Dr Gopal Murty/SPL: 1, 2-3,
Tom Myers/SPL: 28.
David Parker/SPL: 4-5. 34.
Alfred Pasieka/SPL: 6-7, 8, 45, 46-7, 48.
P.Plailly/Eurelios/SPL: 32.
W.A.Ritchie/Roslin Institute/Eurelios/
SPL: 33.
Sinclair Stammers/SPL: 27.
Volger Steger/SPL: 41.

*Every attempt has been made to clear copyright
but should there be any inadvertent omission
please apply to the publisher for rectification.*

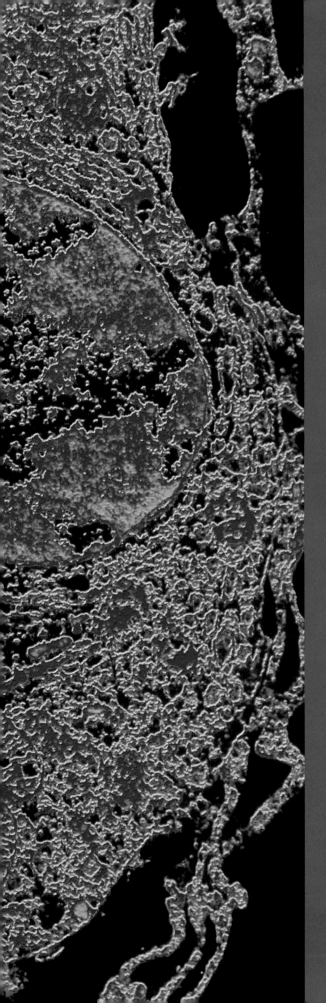

Contents

All About GENETICS

Have you heard about clones, designer babies and animals that grow human body parts? These are all developments in the science of genetics. Extraordinary advances of this kind are announced regularly both on TV and in the newspapers. There's no doubt that in the 21st century you'll hear about the subject more and more. To understand genetics, you need to start understanding what's inside you.

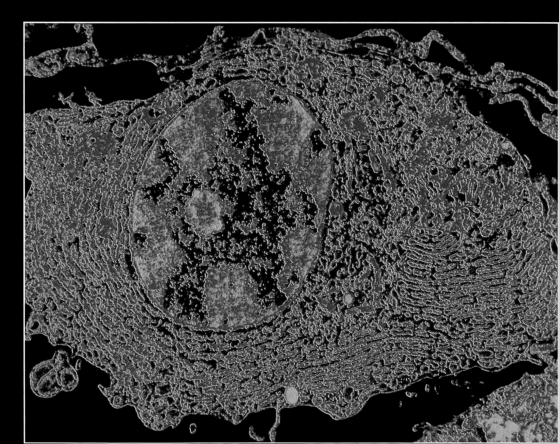

A cross-section of a normal human cell. Genetic material is contained within the cell's nucleus, the round structure in the middle.

Your body is made of tiny cells, thousands of billions of them. Inside each cell there is a nucleus, a kind of control centre that makes the cell work the way it should. Inside each nucleus there are tiny pieces that hold the recipe for how you look and how your body works. The parts of the recipe are called genes; the study of them is called genetics.

In previous centuries nobody knew genes existed. People realised that traits such as red hair or big ears were passed on in families, but no one knew exactly how. Then, during the 20th century, genes were discovered and it became clear that they were responsible for passing on the recipe for life from parents to children. Since then science has surged forward. Now scientists don't just study genes; they can cut them up, remake parts of them differently and even make artificial copies of them.

Controversial knowledge

Scientists of the 21st century are making new genetic breakthroughs, but it's up to all of us how we use their discoveries, for good or bad. So it's important we understand what scientists are talking about when they mention such hot topics as cloning, genetically altered animals and gene therapy, all controversial areas you'll read about in this book. Once you know the facts you'll be better able to join in the debates going on about how the human race is going to use this new knowledge. Perhaps you'll make an incredible new genetics breakthrough of your own one day.

Even if you don't become a scientist yourself, genetics will affect you more and more, almost certainly through the food you eat, the medical care you get and probably in many ways no one has yet even thought of. In this book we make a few guesses about what might happen but genetics moves forward so fast it's often called a 'revolution', so there are likely to be some developments that will surprise us all in the future.

Cells aren't really purple

Human cells are so tiny that 500 of them could fit on the full stop at the end of this sentence. We can only see them by using powerful electron microscopes. The pictures these microscopes show are black and white, but then they are artificially coloured by computers to make them easier to see. This kind of coloured picture is called an electron micrograph, and you'll see some of them in this book.

CHROMOSOMES

To study genes, first find some chromosomes. They are tiny floating rod-shaped objects hidden in each cell nucleus that you have. Each chromosome is a coiled bundle of thread made of material called deoxyribonucleic acid, or DNA for short.

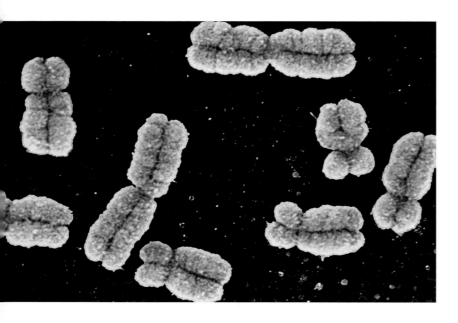

▲
Magnified chromosomes, bunched in pairs.

Virtually all of your cells have the same identical set of chromosomes inside them, carrying a complete DNA recipe for you. If you could unravel the DNA from all the chromosomes in one human cell, it would measure about 1.75 m (5.7 ft) long.

How many chromosomes?

When your life begins you are given 23 chromosomes from your mother and 23 from your father, so nearly all your cells contain 46 chromosomes. The 46 can be arranged into 23 pairs, one half of each pair from your mother and one half from your father.

Geneticists have numbered chromosome pairs to make studying them easier. They are numbered from 1 to 22, with a special pair numbered X and Y that determine whether you are a boy or a girl (you'll find out about these on p 18).

Chromosomes are like tiny packages carrying instructions for a recipe that makes you the way you are and keeps your body operating. The recipe is written along your DNA. Sections of the DNA, different parts of the recipe, are called your genes. Each chromosome has a long length of DNA that carries many different genes.

Occasionally humans are born with a different number of chromosomes, and this causes health difficulties. People with Down's Syndrome are born with an extra copy of chromosome number 21.

What about other species?

All the billions of living things in the world have chromosomes full of DNA in their cells. But the number of chromosomes varies from species to species. For instance, mosquitos only have 6 chromosomes in each cell, but goldfish have 94! However, a lot of the DNA on chromosomes is blank. This is called 'junk' DNA (see p 24). In fact humans carry the most genes of any species.

Chromosome differences

Chromosomes are shaped differently to each other because they contain different amounts of DNA. Some are long and thin; others are short and fat, which helps geneticists identify them.

Each chromosome pair carries a set of genes that no other pair carries. For instance, the instructions for your hair colour might be carried on one pair of chromosomes, whereas the instructions for the size of your nose and the colour of your eyes or hair might be located on the genes within a different pair of chromosomes. Between them your 46 chromosomes carry about 30,000 genes in all.

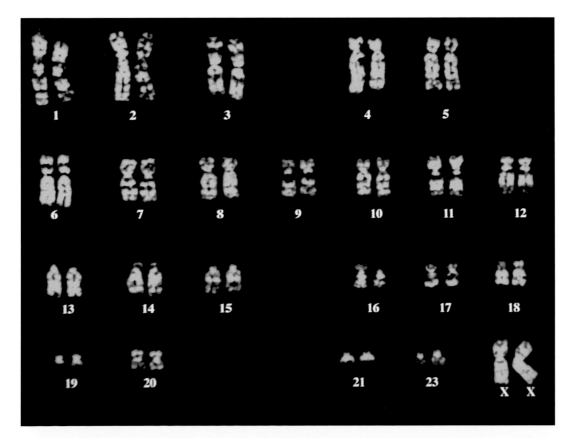

A set of female chromosomes, artificially arranged in pairs.

DNA Your Personal Recipe

DNA is so thin you could fit roughly 5 million DNA threads through the eye of a needle. But under an extremely powerful electron microscope it's possible to see that a DNA thread is in fact two strands with lots of rungs between them, looking like a twisted ladder. This shape is called a 'double helix'. Different sections of the ladder hold different codes for doing specific jobs in the body. These sections are the genes.

The letter code of life

DNA is made of chemicals called nucleotides. There are four different nucleotides, and they join up together to make the DNA ladder. The four chemicals are adenine, thymine, cytosine and guanine. They are given the letters A, T, C and G, which makes it easier for scientists to read the order in which they lie. We can make the order into an alphabet sequence called a genetic code.

Each gene, or section of DNA, has several thousand As, Ts, Cs and Gs joined together in an order different from any other gene.

The nucleotides sit in pairs, one on each side of a ladder rung. And in humans As always pair with Ts, and Cs always pair with Gs. Geneticists call these the 'base pairs' of a gene.

Genes come in lots of different sizes. For instance, the human gene that makes a type of protein called insulin has 1,700 base pairs, or ladder rungs. One of the biggest human genes discovered is one involved in a degenerative disease called muscular dystrophy. This gene is made of about two million base pairs.

What genes do

A genetic code is a recipe for making a protein, a substance that makes your cells the size and shape they are and keeps them operating the way they should. Your DNA contains the genetic codes for thousands of different proteins, all the ones your body needs to grow and survive. Each gene is a recipe code for a different kind of protein.

Life like us

Other lifeforms also have DNA ladders made of As, Ts, Cs and Gs, but they don't have the same genetic codes as those found in humans. After all, their bodies are different to ours, so they need to make at least some body proteins that are different to the ones we need.

In fact human DNA is very different from some species, less so from others. For instance, you share 98.4 per cent of your DNA code with that of a chimpanzee and 90 per cent of it with a mouse. About 33 per cent is the same as a nematode worm and about 1 per cent is the same as a bacterium. However, 99.9 per cent of your DNA is exactly the same as every other human on the planet, which is why we all look roughly the same shape. But that leaves 0.1 per cent of your DNA which is absolutely unique to you, and you alone.

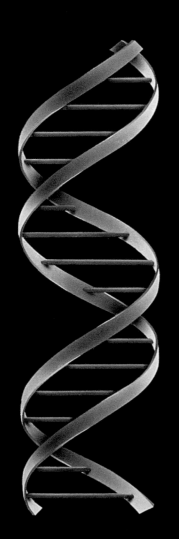

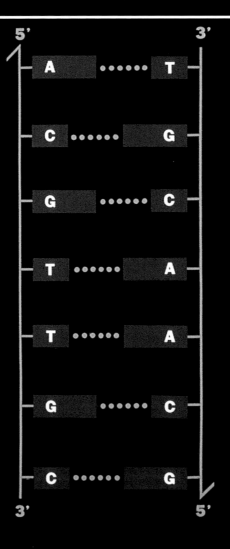

◄◄
An artist's impression of the 'double helix' of a DNA strand.

◄
This diagram shows how nucleotides pair together in a ladder shape.

Genes AT WORK

Before geneticists go about changing genes they need to know how they work. Genes are constantly busy making proteins in your body. It's happening right now, as you read this book.

Know your proteins

There are many thousands of different proteins busy in your body. One obvious one is melanin, a protein that colours your skin and protects it from sun damage. Keratin is the protein that makes your hair and nails, and collagen is a skin protein you might hear about in cosmetics adverts. All your body organs and muscles are made of proteins.

Problems with proteins

Occasionally a genetic code is faulty. This causes the ribosomes – round-shaped protein factories within each cell – to make incorrect proteins that don't do their jobs properly, causing major health problems.

Making proteins

1   A gene carries a code to make a protein. When the body needs this protein the section of DNA that holds the gene unwinds from its chromosome and unzips itself, exposing the gene code.

2   The gene needs new code pieces to make a copy of itself. Luckily there are lots of spare code pieces floating around the nucleus ready to help. They join on to one of the unzipped DNA strands to make a copy strand called 'RNA'.

3   The RNA strand floats out of the nucleus to the outer part of the cell. There it sticks to a ribosome. Each ribosome has lots of protein building blocks called amino acids floating around it. The ribosome reads the genetic code off the RNA and obeys the instructions by gathering the right amino acids to make a particular protein.

4   The amino acids join up in a long string, like a necklace. The necklace folds up tightly and leaves the cell to do its job elsewhere in the body.

A gene with letters missing from its code, or too many letters, can have serious repercussions. For instance, people with Cystic Fibrosis have a faulty gene controlling the mucus production of cells in the lungs. Because the gene is incorrect their lungs become dangerously overclogged with mucus. The majority of Cystic Fibrosis sufferers have three letters missing from a vital gene found on chromosome 7. Since the fault is genetic, scientists hope that one day they will to find a way to cure it, possibly using 'gene therapy' (see p 37).

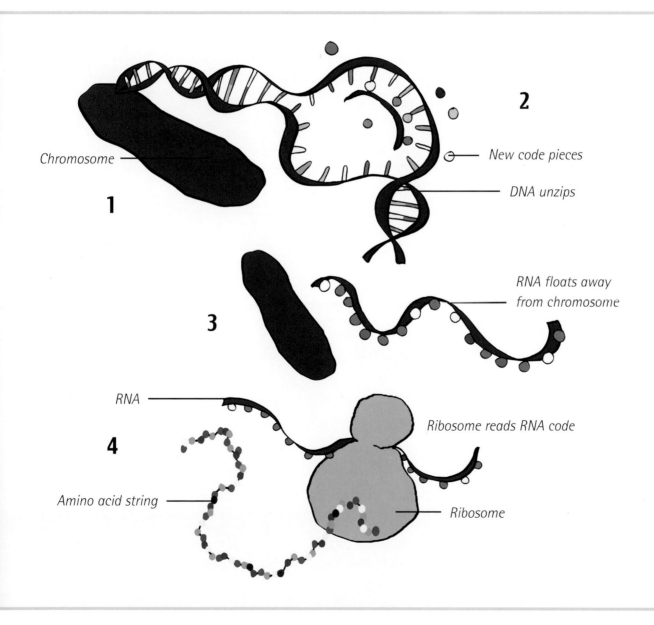

Chromosome

1

2

New code pieces

DNA unzips

3

RNA floats away from chromosome

RNA

4

Ribosome reads RNA code

Amino acid string

Ribosome

Cells make MORE cells

When you were conceived from one egg and one sperm, you began as a single cell. This single cell divided, then the new cells divided again, and so it went on until you eventually became billions of cells, each one with a complete set of chromosomes inside it. This process is called 'mitosis', or the 'cell cycle'.

Mitosis

1 Before a cell divides it takes about eight hours to copy each one of its 46 chromosomes. Then each chromosome bunches up with its new copy. The two together look like a tiny X shape.

2 These X-shaped double chromosomes line up in the middle of the cell. Tiny threads appear, stretching from one side of the cell to the other. The X-shapes attach to the threads.

3 The nucleus membrane, which acts as a thin protective layer, disappears and the cell begins to pull itself apart. The X-shaped double chromosomes pull apart too, so each half ends up on different sides of the cell.

4 The cell pinches together in the middle and breaks into two. Now there are two cells, each with a complete set of 46 chromosomes. The cell division takes about 10 minutes.

1

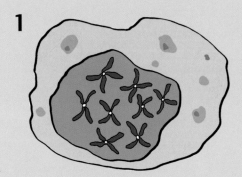

2

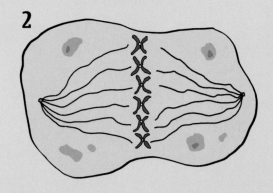

There are 46 chromosomes in a cell, but there is not room to show them all here. Instead we have shown just a few as a representation.

Different cells for different jobs

The first few cells of a new lifeform are all the same. They are called 'stem cells'. As they divide over and over again, they begin to alter to become the many different types of cell a human has. The cells are obeying genetic instructions to alter.

Once a human is fully grown some cells stop dividing. Others continue dividing regularly to replace cells that have died. On average about 100,000 cells are dividing every second inside a human adult, and the same amount die off harmlessly.

Cells out of control

Sometimes a normal cell goes wrong and can't stop copying itself. If the body's immune system doesn't stop the fault, the cell will divide uncontrollably to grow into a cancerous tumour that will begin to harm the other cells around it. As scientists begin to learn more about the genetic prompts that make a cell divide, they move closer towards new cancer cures.

Everyone does it

The genes that prompt cell division are similar in most living creatures. That's helpful to scientists in several ways. For instance, yeast, although a much simpler lifeform than a human, uses proteins very similar to human cells when it divides, so scientists can study yeast experiments to learn more about human-style cell-division. Yeast is much simpler to obtain than human cells and it divides much faster, which makes it more convenient for studying.

Life begins

Nearly all human cells have 46 chromosomes – two of each type. But there are two special ones that are different. These are the egg cells produced by a female and the sperm cells produced by a male. They are called 'sex cells' or 'germ cells'.

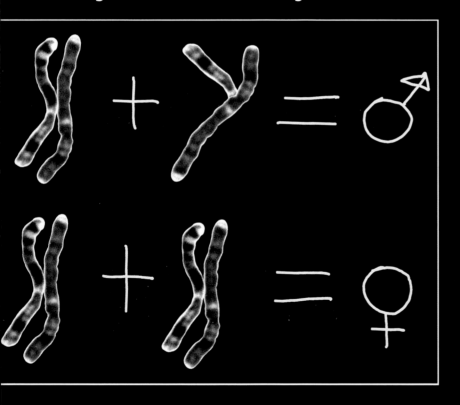

▲

Your chromosome mix decides your gender, either male (top) or female.

divide when they are first made in the body is called meiosis, and it's at this point that an amazing gene swap goes on so that if, in the future, the sex cell is used to produce a child, the child's chromosomes will be very slightly different from its parents.

Boy or girl?

The 23rd chromosome pair in a human cell is different to the others because it varies depending on whether you are a boy or a girl. A female has two equal-sized chromosomes named Xs. A male has one X chromosome and one Y chromosome.

A female egg cell carries one sex chromosome, always a single X. A male sperm cell carries one sex chromosome, but it might be an X or a Y.

At fertilisation egg and sperm join up, and the gender of the child depends on which chromosome the

Sex cells have only 23 chromosomes each. Only when a sperm and an egg join up at conception do they make a full set of 46 chromosomes in the first cell of a new human being.

Sex cells are made differently from normal cells. The way they

Meiosis

1 A cell gets ready to divide, as in mitosis (see p 16). The 46 chromosomes copy themselves and each one joins up with its new copy to make an X-shape.

2 Now each X-shaped double-chromosome finds its matching partner, so there are soon two X-shaped double chromosomes lining up together in the middle of the cell.

3 A unique new gene recipe is about to be made. The matching partners wind around each other and swap some genes between them. It's impossible to predict which swaps will go on.

4 The cell splits apart, as in mitosis. The matching pairs of X-shaped double chromosomes split apart so each new cell has one set.

5 The two new cells are about to divide again. This time the X-shaped double chromosomes split into two halves. The process ends up with four cells, each carrying 23 chromosomes.

sperm cell was carrying.
XX produces a girl.
XY produces a boy.

Choosing gender

It is possible to genetically test the sex of an egg fertilised in a test tube. The fertilised egg can then be implanted in the mother to develop into a new human with a pre-chosen gender. This genetic treatment is sometimes offered to people who are at risk of passing on a serious disease that is only inherited by one gender (see p 23). For instance, a woman at risk of passing on a disease that occurs mostly in boys might choose to have a girl instead.

INHERITING genes

▲

Identical twins share the same genetic codes.

Your genes have been passed down to you by your mother and father. Your parents inherited their genes from their parents, and so on right back to the very first people on the planet.

You received one chromosome copy from each parent. For instance, you got two chromosome 7s, each one carrying the same genes as the other one, for doing exactly the same jobs in the body.

So which gene will your body choose for a particular job, your Dad's or your Mum's? In fact, most of the matching genes on your two chromosome 7s will be exactly the same and it won't matter which one the body uses. But some will have slightly different codes from each other, and in that case one gene code could be dominant over the other. In that case the body will use the dominant gene for the job and ignore the unused 'recessive' gene.

All about eyes

Eye colour is the easiest way to see gene dominance at work because the gene code for brown eyes is always dominant over the gene code for blue eyes. Which eye colour you get depends on how your parents gene mix was shuffled before being passed on to you.

Passing on eye colour

This picture shows representations of parents and their children:

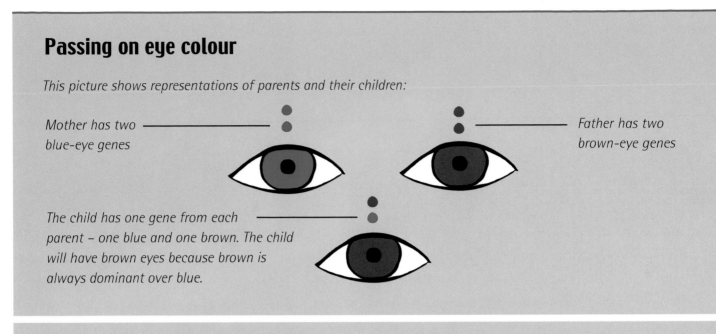

Mother has two blue-eye genes

Father has two brown-eye genes

The child has one gene from each parent – one blue and one brown. The child will have brown eyes because brown is always dominant over blue.

It would be different if the parents' eye genes were more mixed:

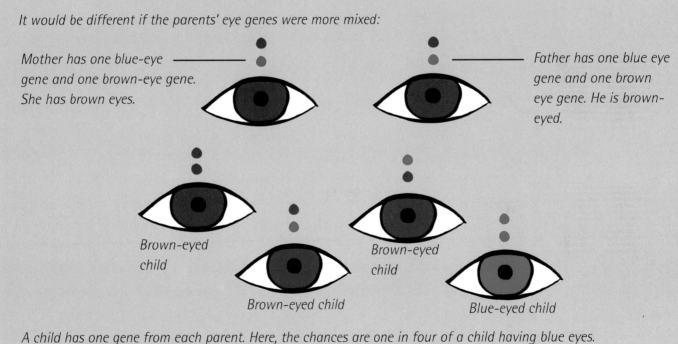

Mother has one blue-eye gene and one brown-eye gene. She has brown eyes.

Father has one blue eye gene and one brown eye gene. He is brown-eyed.

Brown-eyed child

Brown-eyed child

Brown-eyed child

Blue-eyed child

A child has one gene from each parent. Here, the chances are one in four of a child having blue eyes.

Identical twins

Identical twins look exactly the same, from their eye colour to the size of their feet. During their conception one sperm cell joined with one egg cell as usual to make an embryo, with 46 chromosomes inside it. But then the embryo did something unusual. It divided to make two identical embryos. Both developed into humans with exactly the same 46 chromosomes, including the same dominant and recessive gene codes.

When GENES go WRONG

Sometimes people carry a faulty gene code. It's possible they could pass it on to a child in just the same way they might pass on a gene for eye colour. Defects in single genes cause between 3,000 and 4,000 inherited diseases in humans.

Gene faults tend to occur in population groups of similar-type people, because they share a common ancestry. For example, the most common hereditary disease among American people of Northern European descent is Cystic Fibrosis, which affects one US child in every 3,900. One in 31 Americans carries the gene code for Cystic Fibrosis, found on chromosome 7.

By looking at the 'CF' gene area on chromosome 7, geneticists can work out the chances of prospective parents passing on Cystic Fibrosis. Look at the family groups on page 23 to see how this works.

Finding a defect

There are roughly 3.5 billion code letters in a set of human chromosomes, so looking for a defective gene is difficult. Geneticists can make it easier by comparing the gene codes of people with the same health problems and working out how they all differ from a normal gene code.

Arthritis in Iceland

Most Icelanders are descended from a small group of settlers, so they all have very similar DNA. Icelanders tend to suffer from a disease called osteoarthritis, which can cripple them in later life. By looking at lots of Icelandic gene codes scientists have pinpointed a gene fault that helps cause osteoarthritis. In the future, they will be able to test for it and treat people to lessen the chances of it occurring.

Passing on the Cystic Fibrosis gene

Mother has two healthy 'CF' genes

Father has one healthy CF gene and one faulty CF gene.

The father passes on the faulty CF gene, but the mother passes on a healthy CF gene that overrides it.. The child is healthy.

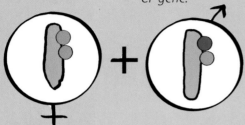 +  =

Mother has one faulty CF gene and one healthy CF gene.

Father has one faulty CF gene and one healthy CF gene.

The resulting child has a one in four chance of receiving two faulty CF genes.

 + 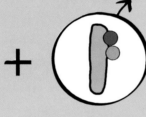 =

Gender problems

Most chromosome pairs have the same set of genes on each half. So if one parent passes on a slightly incorrect code, there is another one to override it. But the sex chromosomes, X and Y, carry very different genetic information to each other. This is why some genetic disorders are likely to be gender-related.

Boys tend to suffer more gender-related disorders than girls. A girl gets two X chromosomes with matching genes so she has some chance of having a correct gene to override a faulty one. But a boy gets an X and a Y carrying sets of very different genes, so there is no chance of getting a correct gene to override a faulty one.

Changing DNA

Cells occasionally make mistakes when they copy their DNA codes to make a new cell. These changes are called 'mutations'. They can be harmless, dangerous or occasionally make a gene better at its job.

If a body cell mutates its gene code the mistake is simply passed on to any new cells it makes in the body. However, if a sex cell – a female's egg or a male's sperm – mutates it could pass the code change on to the next generation if it becomes part of a fertilised embryo.

Gene-changers

Genes can be altered by a few outside influences, such as the sun's radiation or by exposure to some chemicals. These gene-changers are called mutagens. Viruses are important mutagens. These miniscule parasitic lifeforms cannot make proteins themselves; instead they invade other cells and hijack the gene equipment. Some viruses insert their RNA into an invaded cell and use its ribosomes to make their own protein recipes (see p 15). Some viruses, called 'retro-viruses', can actually insert their own DNA material into a host chromosome, which then passes it on when it makes cell copies. Cells have ways of defending themselves against mutations so viruses don't always succeed in their attempt at piracy.

Ghost DNA

In between your genes, the useful coded sections of your DNA, there are long sections of meaningless code letters called 'junk' DNA. Some of it is thought to be 'ghost' DNA, pieces of DNA code that were probably once working genes but no longer function because of mutations that happened over time. It's possible that some junk DNA could be code from ancient viruses that managed to insert their DNA into our ancestors. Junk DNA is not fully understood yet; it's one of the genetic mysteries yet to be solved in the 21st century.

The world's family tree

Genetic mutation is the basis of the theory of evolution. It is suggested that life on earth began as bacteria. Mutagens acted on the bacterial genes to mutate them in different ways, leading gradually over aeons of time to all the different species of life. Harmful mutations led to some species weakening and dying out. Helpful mutations led to some species becoming stronger and better able to survive.

It's possible to map evolution by studying genes common to different species, to work out how closely they are related. For instance, many species share the same general type of cell-dividing genes and it's possible to compare these to see how alike they are. It turns out that the genetic code of a human cell-dividing gene and the code of a chimpanzee cell-dividing gene are very close, so many scientists believe it likely that humans are closely related to chimps. They think that chimps and humans may have come from the same ancestors millions of years ago. Gradually the genes of these ancestors mutated to create separate species.

Changing shape through time

According to the theory of evolution, human skull shape has changed over time due to genetic mutation.

Australopithecus
3.5 million years ago

Homo erectus
500,000 years ago

Homo sapiens
(modern human)

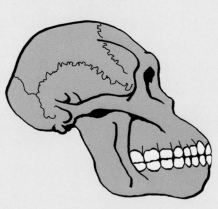
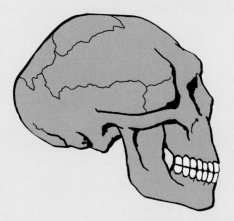
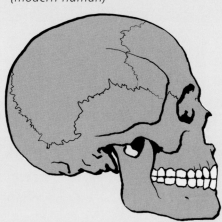

Gene-splicing

DNA unzips and zips itself up again when it makes proteins (see p 15). In order to split apart and join back together it uses chemicals called enzymes that hover around each chromosome, ready to help.

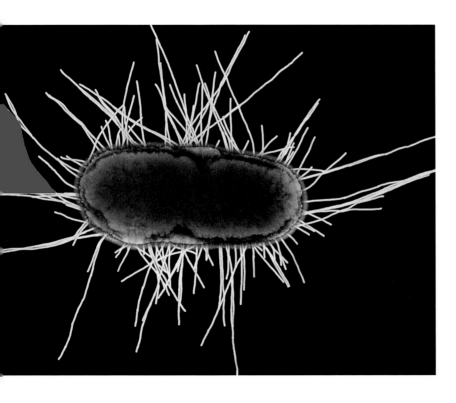

▲

A magnified E. coli *bacterium, relatively easy to alter genetically .*

Some enzymes can snip DNA; others glue it back together. There are over 1,000 cutting enzymes, called 'restriction enzymes', that cut the DNA at different places along the genetic letter code. An enzyme called ligase is the genetic glue that sticks the DNA back together.

Geneticists have found out how to use enzymes as tiny tools in the laboratory to cut up DNA and glue in new genes from other lifeforms to make a version called 'recombinant' DNA. This is called gene-splicing.

Bacteria plus human

Recombinant DNA can make large quantities of medically useful human body proteins. The first step is to find a cell that will replicate itself quickly. *E. coli* bacteria are the best for this. They are relatively easy to genetically alter and they copy themselves very quickly to make new *E. coli* cells.

The next step is to snip the correct protein-making human gene out of the DNA found in an ordinary human cell. The *E. coli* DNA is then snipped open and the human gene is inserted. Kept in the right laboratory conditions a genetically altered *E. coli* will quickly start to reproduce itself, passing on its new DNA code to each new copy made. In a few hours there may be a billion new bacteria, all carrying the human gene and using it to make the protein the scientists want. The protein can be extracted and given to humans who cannot make enough of it themselves.

Putting genetics to work

Genetically altered *E. coli* are used to make insulin, a protein for breaking down blood sugar. People with diabetes lack sufficient insulin and need regular insulin injections. Gene-splicing is also used to produce human growth hormone (to aid body growth) and interferon, a virus-fighting protein. Before the advent of genetic engineering these vital medicines could only be extracted in small quantities from the bodies of mammals such as pigs and cows.

In the food industry bacteria with an added cow-gene provide an enzyme called rennet that starts off the process of turning milk into cheese. Prior to this, rennet could only be taken from calves' stomachs.

In the field of medical research geneticists splice human genes into nematode worms so that new medical drugs can be tested on them before being given to humans.

How far can we go?

The ability to splice genes has some very controversial aspects. Recombinant DNA, splicing genes together from different lifeforms, is now big business and experiments are going on all the time to find useful new gene mixes. But there is fierce debate about who owns the rights to new lifeforms made this way. Should companies be allowed to patent new gene mixtures made using human genes found in all of us? Will harmful new lifeforms be accidentally created, spreading unforeseen diseases? You can find out more about some of the arguments on p 38.

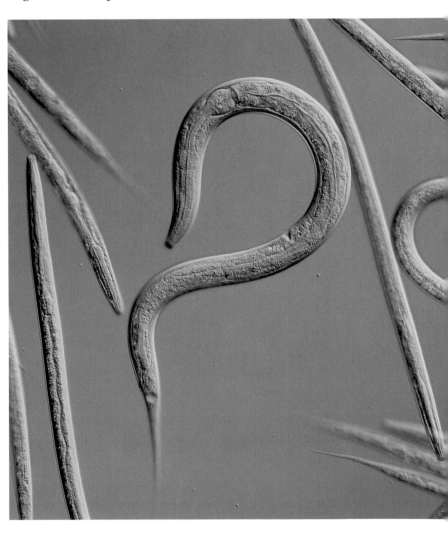

Magnified nematode worms, used for genetic experiments.

PLANT GENETICS

For centuries crop farmers have used genetics naturally by cross-breeding plants and selecting seeds to grow crops that taste good, are resistant to disease and provide a big yield. But genetic technology has changed plant science dramatically. It's now possible to splice a useful gene into a plant to make it resistant to such dangers as disease, attack by insects or cold weather.

Transgenic plants

Plants or animals that contain a gene from another species are called 'transgenic'. There are over 2,000 transgenic plant species, and some of the mixtures that scientists have created are quite surprising. For instance, they have added a flounder fish gene to a tomato, a chicken gene to a potato and a moth gene to a potato. If you ate any of these genetically modified foods (known as GM foods) you wouldn't notice any taste change. But they would have qualities which would make them more commercially successful, perhaps lasting longer or resisting frost damage better. Once again, this science is controversial. You can find out why on p 38.

The first biofood

The first genetically altered food that went on sale was a tomato called a 'FLAVR SAVR ™'. Scientists wanted to create a tomato that would not overripen on its journey from the farm to the shop. They spent $25 million finding the gene that made a firm tomato go soft, and they took it out. Then they reinserted it backwards to cancel out its original effects. In the end it didn't work as well as the scientists hoped, and it was not popular with shoppers. But it was the first of many similar and more successful products.

How to alter a plant

There are some ingenious ways of getting a new gene into a plant cell. One method is to insert the new gene into the DNA of a plant-attacking virus. Then scientists disable the genes that make the virus harmful to the plant. Finally they send it in to invade the plant cell where it inserts its own DNA, including the new gene. Another way is to use a 'gene gun' that fires microscopic metal spheres coated in DNA. Particles of the spheres get sprayed into the plant cell, hopefully inserting some of the DNA. A third way is to give a plant cell an electric shock to make it split open. It will repair itself, but in the meantime new DNA can be inserted.

After a plant cell is genetically altered scientists use chemicals to prompt it to start dividing. Eventually it will grow into a whole new plant carrying the new gene recipe.

Eat your medicine

Edible plants such as bananas, lettuce and tomatoes are being genetically altered to carry disease-resistant vaccines for humans or produce antibodies called 'plantibodies', to prevent virus infections in humans. We can expect lots of new announcements about edible plant genetic breakthroughs in the twenty-first century.

▲

Scientists create seedlings containing a hidden secret – genetically modified DNA.

◄◄

Genetically altered tomatoes were the first biofood to go on sale.

MIXING
ANIMAL DNA

Adding genes from one species of animal into the DNA mix of another produces a 'transgenic animal'. In practice it is not yet straightforward to add new DNA to fertilised animal egg cells. Very often the altered eggs die or fail to incorporate the new DNA. Eggs that survive and begin to divide successfully are implanted into an adult female animal to grow into a transgenic creature.

Why do it?

Transgenic animals are created to make farming more profitable and efficient, for instance to make fatter pigs or sheep that will produce more wool. This business is sometimes referred to as 'pharming'.

Animals are also genetically modified to make human proteins for use in the human body. Sheep, goats or cows with an added human gene will produce milk that contains a human protein which can be extracted, purified and then given to a human unable to produce enough of it naturally. For instance, goats' milk has been produced containing a human anti blood-clotting protein, and by adding human genes scientists hope to make it possible for cows to produce 'humanised

milk' similar to human breast milk.

Transgenics may have uses in the engineering industry, too. To make cabling that is even stronger than steel a spider gene was added to a cow. The resulting cow's milk contained a protein that makes spider web silk. Once extracted and processed this was used to make the super-strong cable.

In medical research fertilised animal egg cells have been genetically altered to make the resulting animal develop diseases found in humans. These lab animals are then used for studying diseases and testing medicines. Very often genetically altered mice are used for this kind of research. These are often referred to as 'designer mice'.

Human organ research

Geneticists hope to use animal organs, such as livers or kidneys, in human bodies. This is called 'xenotransplantation'. Normally a human body would reject an animal part, but if it contained certain human genes rejection might be less likely. Xenotransplantation is still in development. If it does happen, pigs are likely to be the donors because their organs are roughly the same size and shape as human organs.

Body–building

Stem cells are the very first cells a fertilised egg cell makes when it divides. Geneticists have had some success coaxing animal stem cells to grow into specific new organs. For instance, a cow stem cell has been coaxed to grow into a cow kidney in a laboratory jar. These animal stem cell experiments might eventually lead to human body parts being grown in lab conditions from human stem cells.

Is it right?

Using animals in genetics is highly controversial. Many people argue that it infringes animal rights or that it could introduce dangerous animal viruses into the human body. Some suggest there could be unforeseen problems if transgenic animals bred with ordinary animals and new artificially made genetic mixes began to spread unchecked through the animal kingdom. Others answer that the possibility of medical benefits for humans outweighs the ethical issues and that health risks can be avoided if scientists are careful.

▼

A mouse with a gene from a glowing jellyfish added to it so that it glows in the dark.

CLONING

Cloning is one of the most controversial of all new genetic techniques. It is done using a tiny pipette, many times thinner than a human hair, to suck out the nucleus from a female egg cell. Then a new nucleus, taken from another cell, is inserted in its place. After that the egg cell is stimulated with electric pulses to make it start dividing. Each new cell it makes will carry the genetic blueprint from the new nucleus. This method of cloning is called 'nuclear transfer'.

▲

Dolly the sheep – the first cloned animal.

Hello Dolly

Dolly the sheep was born in 1996, the first creature to be produced using nuclear transfer. Embryologist Ian Wilmut removed the nucleus from the egg cell of a Scottish Blackface ewe (a sheep species that has a black face). He inserted a nucleus from a Finn Dorset ewe (a sheep species that is entirely white). The result, Dolly, was white. She had exactly the same genetic material as the Finn Dorset ewe, and none from the Scottish Blackface ewe. Dolly was a clone of the Finn Dorset female: her artificially produced identical twin.

Cloning is by no means straightforward, however. Before Dolly was born Wilmut tried nuclear transfer 277 times. Of these only 29 embryos survived the process, and from this small number only Dolly survived for long. In addition it has become clear that a clone made using a nucleus from the cell of an adult animal can suffer from premature ageing, but it's not fully understood why. There is a lot still to learn about cloning.

Why do it?

Ian Wilmut experimented with Dolly because he hoped to make it easier to create large flocks of genetically altered creatures for medical use, for example, to produce useful human proteins in milk. Other scientists have been involved in cloning endangered animal species. For instance, Noah the guar, one of a rare cattle-related species, was cloned by putting a guar nucleus into a cow's egg cell, then implanting it into a cow's womb to grow into a guar baby.

It's possible that other endangered creatures such as giant pandas or tigers will be cloned, as well as beloved pets. It may even be possible to clone extinct species from DNA surviving in body remains. For instance, scientists have attempted to clone a woolly mammoth from DNA frozen in ice thousands of years ago, so far without success because the DNA may be too old and damaged. Human cloning could theoretically help infertile couples to have a child.

Human cloning

As yet, a human clone has not yet been produced. Nuclear transfer has been conducted on human egg cells and the results have divided into a few new cells, but the process has not been allowed to go further.

To clone a human you would first need an egg cell donated by a female. You would then need to replace its nucleus with one from another person, and stimulate it to grow, then implant it in a human female womb to develop into a baby. As yet there would be no guarantee of success since the chances of miscarriage would be very high. If, one day, science progresses to make it possible, should it be allowed? Read some of the arguments on p 38.

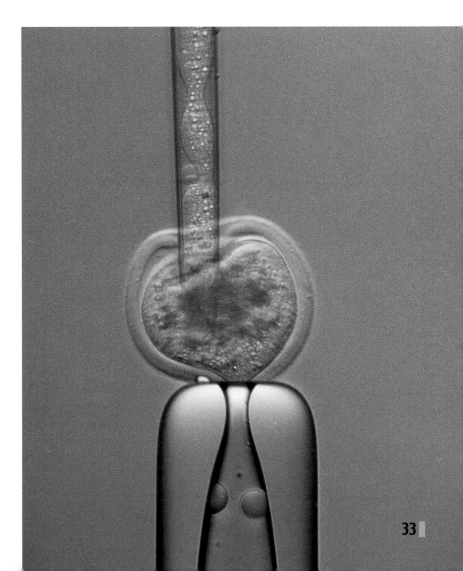

In nuclear transfer, a nucleus is sucked out of a cell using a tiny pipette.

Is that YOUR DNA?

Everyone has some unique sections of DNA code and it's possible to read them, a technique sometimes called 'DNA fingerprinting'. It's used to solve crimes by comparing a suspect's DNA to some DNA material taken from a crime scene, perhaps a drop of blood or a single hair.

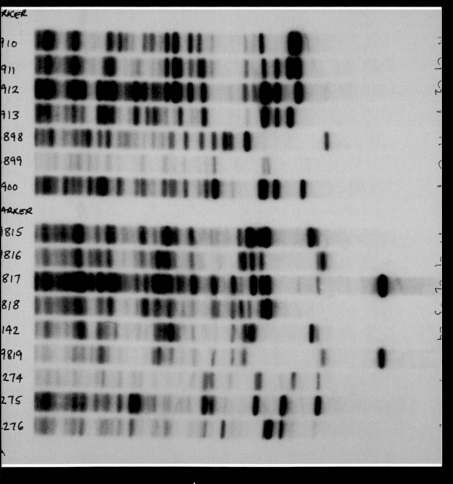

▲ *A DNA strip reveals a unique pattern of bars known as a 'DNA fingerprint'.*

The easiest way to find unique stretches of DNA is to look at the non-coded 'junk' sections of DNA between the genes. There are lots of letter repetitions here, which are unique to one individual.

How to read DNA

Geneticists can make a DNA strip that looks like a barcode. They can compare the pattern of bars on the strip to a strip made from another DNA sample. If it matches, the chances are very high that both DNA samples came from the same person.

To do this, first some DNA is extracted from a cell. Then an enzyme is added that cuts the DNA into pieces at specific points in its letter code. The pieces arrange themselves into bands of different sizes which are then marked to make them radioactive. An X-ray film picks up

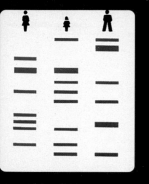

A positive DNA test confirms the identities of both parents of a child.

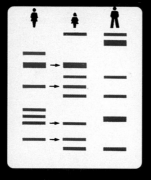

Here, the mother's DNA bars match the child's DNA bars...

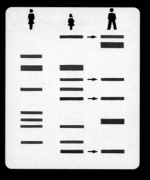

...and here the father's DNA bars match the child's DNA bars.

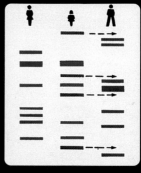

This test shows the correct mother but the wrong father.

the radiation and the result is a picture of the DNA bands.

The probability of finding someone with exactly the same DNA picture as you from exactly the same stretch of DNA as you is about 1 in 10 billion. To make even more certain of a match geneticists compare several strips made from different sections.

Mums and Dads

You share some of your DNA code with your parents, so by using DNA testing it's possible to pinpoint your mother, father and close relatives. Relatives will have a number of matching bands on their DNA 'barcodes'.

One of the most famous recent cases of DNA analysis concerned a woman called Anna Anderson who claimed to be Anastasia, the daughter of the last Czar of Russia. The Czar and his family were killed at the beginning of the Russian Revolution, but Anna claimed she had escaped and had rights to the family fortune. After her death some of her hair was tested and compared to DNA from the bones of the dead Czar's family. Her DNA did not match up to theirs.

Relatives from history

It's sometimes possible to extract DNA from ancient remains and then compare it to modern humans. This happened in 1997, when DNA was taken from the tooth of a caveman who lived in Cheddar, Somerset, about 9,000 years ago. Tests on modern local people revealed that a teacher called Adrian Targett was a relative of the caveman. He still lived in Cheddar, close to the cave where the ancient bones were found.

Some of our DNA can be tracked back even further, to very early humans who lived in Africa.

medical DNA

In the 21st century advances in medical care will be helped along by the completion of the Human Genome Project, the record of the code of every human gene.

Making the record was a huge task because there were about three billion genetic code letters to read, but now geneticists know all the gene codes and where to find them on our chromosomes. That makes it easier to pinpoint code errors and perhaps create medicines to treat harmful ones. In theory it is possible to correct a faulty human gene or add a healthy one that is missing. This is called 'gene therapy'. In practice we are still only at the beginning of the research. One major hurdle still to overcome is understanding how different genes act together. Without knowing that, it's possible that some genetic engineering might cause more harm than good. This is one of the arguments scientists use to justify experimenting on other animal species first.

Fixing harmful genes

If a faulty gene is making a harmful kind of protein it may be possible to put in a new engineered gene to counteract the harmful one. Scientists have created a synthetic molecule of DNA, called a 'chimera molecule', which can enter a cell. It could eventually be used to fix mutations.

Researchers are also working on creating artificial chromosomes by joining lots of DNA together in a laboratory. This might be useful for a patient who needs lots of gene changes rather than just one.

Vaccines and disease

Vaccines against diseases are usually made from a weakened strain of the disease itself. The vaccine is injected into the body to make the body's immune system fight the disease, in the process creating proteins called antibodies that will always protect against the virus in the future. Unfortunately vaccines sometimes have side-effects on the rest of the body.

Adding a missing gene

The first successful gene therapy was done on a little girl called Ashanti DeSilva in 1990. She lacked a gene critical to her immune system, so she had no defence against germs and faced a life in a sealed germ-free environment. The missing gene created a vital enzyme needed in Ashanti's bloodstream.

Ashanti's doctors inserted the missing gene into a virus that could get into human cells and insert its DNA. They disarmed the virus's harmful properties; then they added it to some of Ashanti's blood cells. It got into the nucleus of each one and added the new gene. When the blood cells were put back into Ashanti, her body began to make the missing enzyme.

By using gene technology on mice scientists have been able to produce supplies of antibodies that target specific disease invaders without side-effects. They are nicknamed 'magic bullets' because they go straight to the target, without causing problems elsewhere in the body. It's possible that versions of this highly efficient medical treatment may one day be useful in fighting such diseases as cancer.

Genetic material is stored in a laboratory to help scientists develop new gene therapies.

HOW WILL WE USE GENETICS?

Like it or not, lots of genetic research is going on and we are gradually learning more and more. In the 21st century the big question will be how do we use the knowledge? Here are some of the arguments for and against new developments and future possibilities.

Using humans

In theory it is possible to alter the genetic make-up of a fertilised human egg cell before it develops into a human body. This is called 'germ-line therapy' and it is illegal in many countries because of the uncertainties of what might result or because of religious beliefs. Some people argue that from the moment of fertilisation an egg is a human, and shouldn't be touched. Others believe that until the egg begins to develop into body cells it is not fully a human, just a kind of biological building block that could be used for study. In the course of research on human egg cells there are likely to be many failures. To some this would be tantamount to murder. In the meantime germ-line therapy experiments are carried out on animal egg cells, which upsets animal-rights campaigners.

Do you want to know?

But advances in gene science have both advantages and drawbacks. For example, imagine you discover that you are carrying a gene which could cause a serious disease in later life. On the one hand perhaps you could have treatment to prevent the disease, and change your lifestyle to try to avoid it.

On the other hand you might face discrimination from others if they knew you carried the gene. You might be unable to get insurance or perhaps even a job. You would also have to face the probability of serious illness.

Genetics in the environment

Some people argue that genetically altered animals and plants might mix with 'normal' versions, and eventually pass on their new genes to a whole species. Because GM (genetically modified) crops are grown outside it is possible they could pollinate ordinary crops nearby and pass on their gene code. The new genes could eventually spread into the whole plant species with unforeseen effects. GM crop-testing is going on, but in some countries GM crops are banned.

Some commercial seed companies have genetically altered crop plants to make them sterile, which means they no longer produce seeds which can be gathered and planted the following year to grow a new crop, the traditional practice in countries such as Africa. Instead farmers are forced to buy new seeds each year. If this kind of GM crop crosses with ordinary plants, the result could be disaster and famine on a wide scale.

Owning genes

Science companies have obtained patents on genetically altered plant species, animal species and even human cells. This means they can't be used in any way without a payment to the company which owns them. On the one hand this means that scientists make a profit from their hard work. On the other hand, commercial interests gain a monopoly on forms of life.

▼

GM plants produce high-yield, disease-resistant crops, but could they spread unchecked?

GENETIC
MYSTERIES

Some scientists have searched, so far without success, for genes that lead to criminal behaviour or genes that lead to people making different sexual choices. It's not yet proven whether any behaviour is strongly influenced by genetics.

Do genes win medals?

Lots of research is going on to discover performance genes, to find out whether the world's best athletes have special genes that make them better sportspeople. So far, one gene, called the ACE gene, has been linked with exceptional endurance helpful in such sports as rowing and long-distance running. People with a long ACE gene seem to have hearts that work more effectively during endurance tests, compared to people with a shorter ACE gene.

If performance genes are pinpointed perhaps young people with such genes will be selected and trained to win medals. However, it may be the training itself that proves to be much more important than genetic make-up.

The way ahead or just a side-road?

Some people believe that the importance of genetics has been exaggerated, and in reality such extreme ideas as human cloning and gene-splicing into humans are likely to be impossible. One thing is certainly true; we still have a lot to learn about genetics .

▼
Genetics will continue to be one of the fastest-moving areas of science in the 21st century.

457

458

GLOSSARY

adenine
One of the four chemicals that make up DNA.

amino acids
Chemicals that join together to make proteins.

base pair
A pair of chemicals that make up one rung of a DNA ladder.

chromosome
A coiled bundle of DNA floating inside a cell nucleus.

cloning
Copying an organism by replacing the nucleus of a female egg cell.

chimera molecule
A synthetic molecule of DNA made by scientists.

Cystic Fibrosis
A serious lung disease caused by an incorrect gene coding.

cytosine
One of the four chemicals that make up DNA.

DNA
Long strand of genetic information made of deoxyribonucleic acid

DNA fingerprint
A visual pattern of bars obtained from a length of DNA.

dominant gene
A gene that the body will always use, given a choice between genes.

double helix
The twisted ladder shape of a DNA strand.

Down's Syndrome
A medical condition caused by having an extra chromosome 21.

electron micrograph
An artificially coloured electron microscope photograph.

embryo
A female egg cell fertilised with a male sperm cell.

enzymes
Naturally produced chemicals used to do jobs in the body.

evolution
The theory that lifeforms have altered over time.

gender
Male or female.

gene
A section of DNA strand containing a code for making a specific protein.

gene-splicing
Slicing DNA apart and inserting a new piece.

gene therapy
Correcting a faulty gene or adding a missing gene for medical reasons.

genome
The complete genetic code of a species.

germ-line therapy
Altering genes in a fertilised human egg cell before it develops further.

GM
Abbreviation for 'genetically modified'.

guanine
One of the four chemicals that make up DNA.

human genome project
A record of the code of every human gene.

insulin
A protein used in the body.

junk DNA
DNA containing jumbled coding.

keratin
A protein that makes hair and nails.

ligase
An enzyme that can stick DNA back together.

meiosis
The division of chromosomes that occurs when a sex cell is made.

melanin
A protein that colours the skin and protects it from the sun.

mitosis
The process of cell division.

mutagens
Outside influences that can alter a cell's genetic code.

mutation
When a cell makes a mistake copying its gene code and passes on the mistake to a new cell.

nucleotides
Four different chemicals that join in pairs to make a DNA strand.

nucleus
The control centre inside a cell.

recessive gene
A gene that the body will ignore if a dominant version is present.

recombinant DNA
DNA with new genes added in.

RNA strand
A copy of a section of DNA coding.

restriction enzyme
An enzyme that can cut DNA.

retrovirus
Virus capable of inserting its own DNA into another organism.

ribosome
Protein-making apparatus in a cell.

sex cells
A female egg cell or a male sperm cell.

stem cell
A general-purpose body cell.

transgenic species
A plant or animal containing a gene from another species.

thymine
One of the four chemicals that make up DNA.

tumour
Clump of cells occuring when a cell begins to divide uncontrollably.

X chromosome
Female sex chromosome

Y chromosome
Male sex chromosome

xenotransplantation
Transplanting animal organs into human bodies.

21ˢᵗ century SCIENCE

INDEX

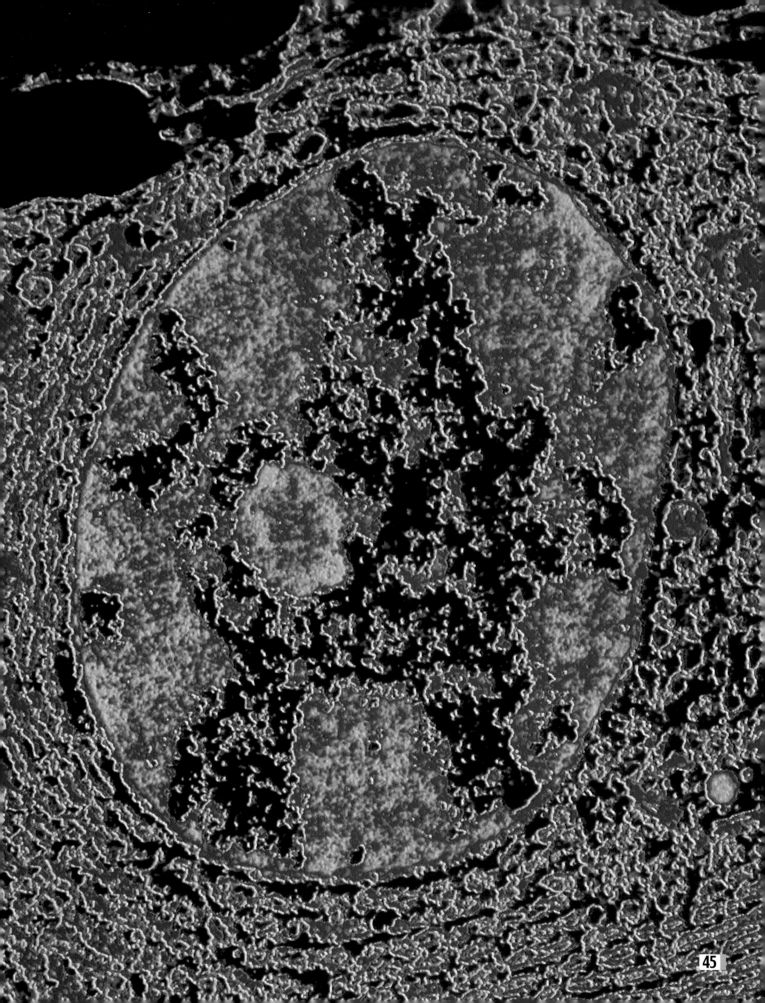

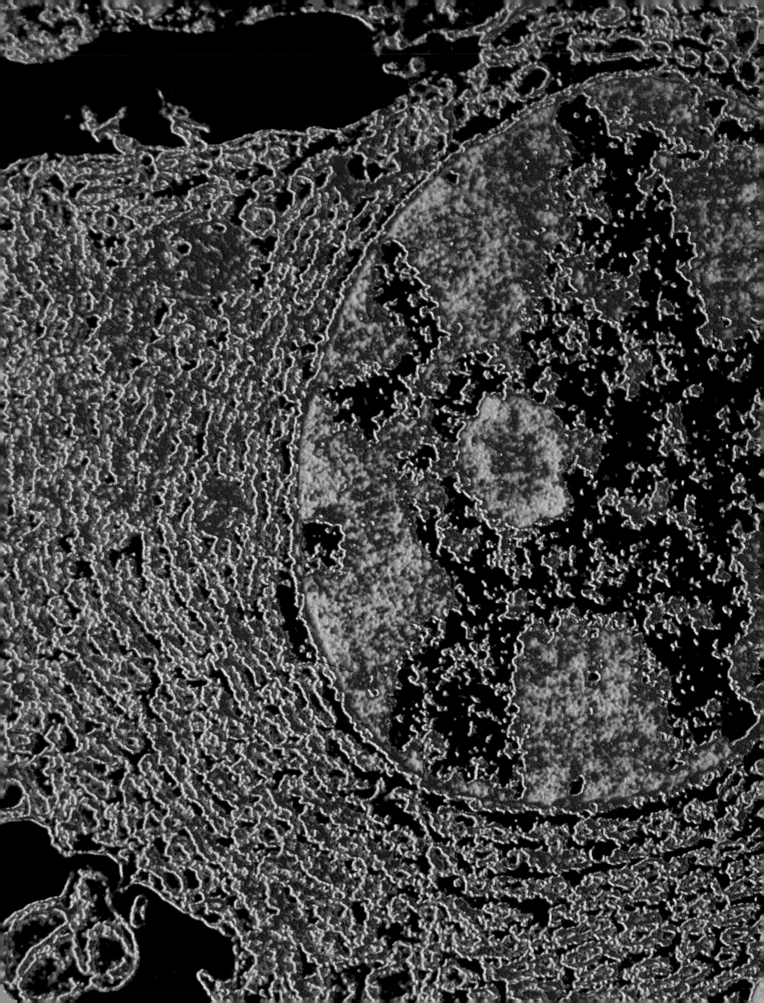